The Fungal Flora of Mull—Additions

Roy Watling

Abstract

A commentary on the fungal flora of the island of Mull, with particular emphasis on the macrofungi, introduces a systematically arranged list of additions to the fungal flora. This fungal flora of Mull has been considerably enriched by nearly three hundred records of rare or previously unrecorded taxa including several new to science.

ISBN 0 9504270 3 9

Introduction

It is now well over eight years since the materials were drawn together for the initial lists of taxa for the fungi of Mull in Jermy & Crabbe (Henderson & Watling, in The Island of Mull. British Museum, 1978). Inevitably several collections were not included in the published work because of lack of information on their identity and because they required further study and even formal diagnosis. To these have to be added the results from an examination of the contents of boxes of undetermined material from various sources and new records particularly those freely offered by Malcolm C. and Marjorie E. Clark, made on their regular holidays on the island. The compilation is therefore a result of the drawing together of all this additional information. It is evident that this second list is fairly extensive and the number of records previously based on single collections has been considerably reduced. Many records are also new to the Hebridean flora, and several taxa recorded from the Western Isles appear to be rather rare elsewhere in the British Isles.

The Mull fungal flora should not be considered in isolation, nor necessarily thought of as any more extensive than the floras of neighbouring islands. Dennis & Watling (1983) have indicated that basically our knowledge of the mycoflora of this province of north-west Scotland is at present almost entirely based on accounts from Rhum (906) and Mull (2007) but our knowledge of the whole Hebridean flora is rapidly expanding. Surprisingly the fungi of Skye, an island of comparable size to Mull and in the same island system, have been largely neglected in the past; this is now being rectified (Watling, 1983b), along with a documentation of the fungi of some of the smaller islands not previously studied, e.g. Jura and Islay (Watling, 1983a).

It is now possible, therefore, to attempt a few comparisons between the mycofloras of Mull and Skye. The taxa recorded for Skye (203), but not for Mull, are by their nature doubtfully unique to Skye, and almost certainly will ultimately be found on Mull. The presence of these species indicates that the Mull flora is still not yet exhaustively recorded. The total number of fungi known from the Inner Hebrides is a little over 2,400 taxa; an additional c. 225 taxa have been recorded for the Outer Hebrides.

Dennis & Watling (1983) have analysed the flora of the more conspicuous larger fungi of Mull and compared it with the appropriate parts of the floras of SE England (Dennis, 1973), Warwickshire (Clark, 1980) and Yorkshire (updated Massee and Crossland, 1905—Watling, unpublished data)—three comparatively well-known floras. These comparisons can be extended only a little further with the paucity of our present data but a numerical comparison between the fungi of the various islands is being made by Dennis (pers. comm.). It is now also possible to make comparisons with Arran, an island in the Firth of Clyde (Kirk & Spooner, 1984) as an extensive fungal survey there has been completed.

FUNGAL COMMUNITIES

Fungal diversity is undoubtedly very dependent on the accompanying vascular plant flora and when this is rich, as on the larger islands of the Inner Hebrides, the fungi are correspondingly more varied. Floristically the Hebrides simply appear to be an isolated part of the general vegetation characteristic of north-western Scotland (McVean & Ratcliffe, 1962). However, although the vegetation of the larger islands differs little from similar habitats on the mainland, the small islands are to varying degrees affected by salt-spray and exposure. It is not considered that there are any really insurmountable barriers to fungal colonization between the islands even with the furthermost islands in the St Kilda group.

In minutiae the islands do not apparently form a natural vegetational unit and no easy separation, agreeing with the topographical division into the Outer and Inner Hebrides, can be achieved. In general terms the Long Isle, Coll, Tiree etc, Skye and Rhum can be considered part of the north Highland unit of 'predominant birch forest' (McVean, 1964) while the remainder including Mull, Colonsay and Oronsay must be related to the west central region of 'predominant oak forest with birch' (McVean, 1964). The specialized habitats which particularly characterize the Hebrides are: 1. herb-rich, machair grasslands; 2. vegetation characteristic of warm, damp Atlantic areas.

MOORLAND

The Uists, Barra and Lewis have large areas dominated by *Calluna-Eriophorum* and *Trichophorum-Eriophorum* communities; sizeable areas of the same bog-vegetation are found on Mull and Skye, and on Jura and Islay, but mycologically such vegetation is rather depauperate, not offering the diversity of habitats as do the island shrub or tree communities. *Molinia-Calluna* moor is characteristic throughout the islands from St Kilda even to Arran, and in some cases *Calluna* can be found in almost pure stands.

GRASSLAND

Agrostis-Festuca grasslands are much richer in larger fungi than the heaths and are especially characterized by members of the two agaric families Hygrophoraceae and Entolomataceae (Watling, 1962). These grasslands are best seen on the basalt slopes of Mull and Skye and if attempts are made to eliminate grazing then the vegetation becomes rich in herb species with a corresponding fall in the magnitude of the fruiting of agarics. This has been demonstrated dramatically on Rhum in the Fionchra area, where experimental plots have been set-up to monitor the development of herbaceous plants within the sward. Whereas moderate grazing often stimulates fruiting, perhaps by in-put of fertilizer, the total elimination of grazing for a three month period was sufficient to allow the development of grass and forbs to become luxuriant enough to suppress extensive production of fruiting structures. Over-grazing and the associated trampling also tends to reduce production of agarics to a state equal to where grasses go unchecked. The degree of animal grazing therefore is an important parameter in discouraging the appearance of agaric fructifications. Outside the base-rich areas the *Agrostis-Festuca* grasslands on Mull often only exist as pockets within the *Calluna* communities and are termed 'greens'; their presence reflects an increase in the base-status of the soil and apparently this encourages agaric colonization, if fruiting can be taken as a measure. However, because the grasses in these 'greens' are less fibrous than *Molinia* or *Nardus,* sheep and deer are attracted to them and trampling becomes a limiting feature tending to reduce fruiting.

Trampling by sheep and deer is found in the sub-montane grasslands, e.g. Ben More, which often contain abundant quantities of the palatable *Anthoxanthum odoratum* or *Deschampsia flexuosa.* Generally, however, such grasslands, although richer than the surrounding Callunetum, are not as rich in agarics as the grasslands at low altitude. In fact typical grassland species such as *Hygrocybe intermedia* and *H. citrinovirens* are replaced by *H. punicea* and *H. laeta* in these communities. The low alpine pastures with *Nardus* found on Skye, Rhum and the central massive of Mull and characterized by *H. coccineocrenata,* are generally disappointing for agarics.

The *Agrostis-Festuca* maritime grasslands with their development of *Festuca rubra* and the machair are perhaps the richest grassland communities. Members of the Agaricaceae are common on these popular sheep grazed lawns, and the local crofters have long collected 'field and horse mushrooms' for the pot. The machair grasslands are characteristic of the Hebridean scene, particularly Coll and Tiree, but are not a feature of Mull.

It has become possible to recognize a rather rich fungal flora for the maritime grasslands (Watling & Richardson, 1970) and it includes agarics in the Hygrophoraceae and Entolomataceae, and clitocyboid members of the Tricholomataceae; amongst the other basidiomycetes the puff-balls (Gasteromycetes) are common, as are the earth-tongues (Geoglossaceae: ascomycetes). In such communities *Hygrocybe langei, H. berkeleyi, Calocybe carnea* and *Agaricus macrosporus* are frequent and go to make up a flora comparable to the grassland communities on limestone pavement in North Yorkshire. Where the sand has been blown on to peaty soils and *Agrostis tenuis* becomes dominant, and even *Calluna* invades the community, *H. punicea, H. laeta* and *H. marchii* replace the broader spectrum of Hygrophoraceae; parallels are found in a reduction of members of the Entolomataceae.

MOUNTAIN COMMUNITIES

Montane vegetation consisting of *Rhacomitrium* heath, or plant-cover associated with erosion surfaces are confined to Ben More in Mull, the Rhum heights and the Cuillins of Skye. Although depauperate in agarics these communities support some of our most interesting species, e.g. *Hygrocybe lilacina* with *Rhacomitrium* on Ben More.

Omphalina luteovitellina is common on peaty soil in mainland Scotland growing from 600m above sea level upwards and is associated with the lichen *Botrydina vulgaris*. In the islands it is also common but can be found at much lower elevations—a shift in altitudinal distribution in keeping with many vascular plants. *O. hudsoniana* (=*O. luteolilacina*) shows a similar pattern of distribution but this species is associated not with *Botrydina* but with the lichen *Coriscium viride*. Montane *Carex-Saxifraga aiziodes* flushes and similar communities, where calcareous drainage water seeps out of glacial drift, are widespread in the islands and support, not only a rich vascular plant flora, but in the drier areas, a rich agaric flora. Such base rich communities contrast markedly with the rather depauperate vegetation on peaty soils and podsols which often surround them, and which are dominated by species of *Galerina* and *Hypholoma*.

In Scottish montane communities, a marked increase in the number of larger fungi fruiting generally occurs where the grasses and forbs are joined by dwarf willows (*Salix* spp.), although this has yet to be demonstrated to any large degree on Mull. The roots of these montane willows (*S. herbacea, S. reticulata*) are associated with fungi parallel in all ways to those on the roots of the dominant trees of a lowland silvicolous community, except where the willows are found in sites with thin soils, e.g. on rock-ledges. Under these unfavourable circumstances the roots are clothed in a pseudomycorrhiza of *Cenococcum granuliforme,* probably because this species is more adapted to surviving lengthy periods of drying out, conditions characterizing these often very exposed places.

WOODLANDS

Oak (*Quercus*) and hazel (*Corylus*) scrub are important components of the Mull flora, and can be also found on Arran, Islay, Eigg, Raasay and Skye, and the fungi of these communities contributed significant numbers of taxa to the Mull list (Henderson & Watling, 1978) with characteristic species of *Russula* and *Lactarius* being recorded, e.g. *Lactarius volemus* in oakwood.

Reafforestation is an important feature of mainland Scotland with pines (*Pinus* spp.) and spruces (*Picea* spp.) being planted widely; a few less familiar species are, in addition, grown in the west. It is also a major factor in the economy of Mull and large areas have been planted. Associated with these conifers are many agarics and related fungi but many are doubtfully members of our native flora. Some of these same fungi are recorded for the islands being associated there with groups or even single trees, e.g. *Suillus grevillei* (=*Boletus elegans* of old texts) with larch (*Larix*) and correspondingly the agaric-flora of such areas is very rich.

Watling & Richardson (1971) described from St Kilda a community of agarics associated with *Luzula sylvatica* (wood-rush). One constituent of this community is *Nolanea cetrata* (Entolomataceae) a woodland species most commonly found under conifers on mainland Scotland particularly in plantations; it is common in conifer plantings on Mull. Dennis (1952) also recorded this species from Ben Loyal, Sutherland where no conifers were to be seen and in parallel the present author has found the same *Luzula* and *Nolanea* on Ben More, Mull. It is tempting to suggest that the *Luzula* along with *Nolanea cetrata* is a relic of former woodland. Or does it indicate that the climate of the mountains of northern Scotland and Mull and the exposed island of St Kilda is parallel to the microclimate associated with Scottish lowland woods in the way that similar ideas have been sought in explaining the European distribution of holly (*Ilex aquifolium*)? All the above observations suggest that whenever plant-communities are considered in the future, larger fungi could offer useful additional information and should not be ignored.

Crofting is the primary land use in the Hebridean islands; a mixture of agriculture and cattle and sheep grazing is traditional, and reclamation of moorland has been carried out in various places. It is estimated that about 22% of the vegetation of the Hebrides has been modified in this way (Currie, 1979; Currie & Murray, 1983). Fertilizing and reseeding has improved grassland with a corresponding increase in the base-status of the soil. Any moves like this, increase the diversity of agarics found in the treated hill-pasture when compared with the original moorland. There is little doubt that when sufficient information is obtained on the floras of larger fungi for individual islands, and specific sites on these islands, that some idea of the history of the communities can be deduced, as previously outlined for plant-cover containing creeping willow (*Salix repens*) by Watling (1981) or for communities on Rhum (Watling, 1970).

HOST-RELATIONSHIPS: BASIDIOMYCETES

It is very interesting to compare the percentages of life-forms of larger fungi for Mull as Watling (1970) attempted for the island of Rhum. These earlier figures were based on his own observations and Dennis' analysis (Dennis, 1964) resulting from the comparison of Rhum, the Canary Islands and the Faeroes. Thus, mycorrhizal agarics and rusts can be taken as subjects for analysis as they are relatively well-known. Although Skye is under-collected the overall picture demonstrated by Dennis (1964, etc) for the Hebridean rusts is maintained (Table 3).

Mycorrhizal fungi are widespread in the Hebrides even being found on Hirta in the St Kilda group where they grow in association with *Salix repens* (Watling & Richardson, 1971). The wealth of numbers of taxa on Mull and Skye can be explained in part by the extensive planting of Pinaceae as mentioned above, and with which agarics are mycorrhizal (Table 1).

Even small plantings of exotic trees appear disproportionally rich in the number of species they support; similar patterns have been found in the fungal flora of the Faeroes (Moeller, 1945) undoubtedly associated with the few plantings of conifers there (Table 2).

Normally host-specificity is considered to be a feature of the larger mycorrhizal fungi, but with more extensive data to hand it is now less convincing as a characteristic of this group of fungi than previously

TABLE 1. Comparison of agaric floras of Rhum, Mull, Skye and St Kilda.

	Rhum	Mull	Skye	St Kilda
Suspected mycorrhizal agarics and boleti	110	282*	84	11
Non-mycorrhizal agarics	178	332	146	151
TOTAL	288	614	230	162

Cyphellaceous fungi, although reduced agarics, are not included in the calculations.
*The large number recorded for Mull reflects the extensive areas planted with native and exotic trees.

TABLE 2. Number of Faeroese agarics in floras of Rhum, Mull, Skye and St Kilda

	Rhum	Mull	Skye	St Kilda	Faeroes
Mycorrhizal agarics	6	13	4	2	21
Non-mycorrhizal agarics	46	60	35	48	121
TOTAL	52	73	39	50	142
Total No. of Agarics	288	614	230	162	142

thought. As records are accumulated other host-fungus relationships are found and although less widespread, none the less important in any one selected plant community. Thus, although usually found with one particular tree, some fungi, e.g. *Suillus grevillei* with larch (*Larix* spp.), *Lactarius rufus* with Scots pine (*Pinus sylvestris*) and *Amanita muscaria* with birch (*Betula*), occur with other tree hosts; e.g. *Lactarius rufus* with *Betula*, *Amanita muscaria* and *L. rufus* with *Picea sitchensis* (Sitka spruce): see *Mull Flora:* 15. 57 (1978). At least one species recorded herein, the Australian hypogeous fungus *Hydnangium carneum,* has undoubtedly followed its host to Mull, possibly in the soil around the roots (or on the roots themselves) of the *Eucalyptus.*

Rusts are also very host-specific and therefore depend for their inclusion in the flora on the presence of their host or hosts. This group of fungi therefore can be used as a measure of that flora. The overall picture described by Dennis (1964) for the rusts of Rhum is supported by the flora of Mull, and even by the rather under-collected flora of Skye (Table 3).

With the saprotrophic agarics and polypores the limitations to distribution are rather different, it being necessary to have not only the host but

TABLE 3. Comparison of Rusts of the Ebudes with Iceland (Jørstad, 1958), Faeroes (Moeller, 1945) and the Canary Is. (Jørstad, 1952). Rhum and Long Is. from Dennis (1964, 1975). Acknowledgements to A. Bennell.

	Skye	Mull	Rhum	Long Is	Iceland	Faeroes	Canary Is.	% of forms within Hebridean flora
Micro-forms	13	13	4	9	16	6	6	19.8% (20)
Lepto-forms	1	4	1	0	1	—	7	1% (1)
Brachy-forms	5	7	3	12	2	3	7	5.9% (6)
Autoecious forms	17	27	6	15	10	9	24	32.6% (34)
Heteroecious forms	22	30	18	20	19	16	30	33.6% (34)
Uncertain	4	4	3	—	3	2	8	6.9% (7)
TOTAL	62	85	35	56	51	36	82	

Total number of Ebudes species: 111.

suitable parts of that host available, and even the host in a particular state of decay. For example *Hypholoma fasciculare* usually grows on hardwoods and is less frequently found on conifer stumps, whilst *Tricholomopsis rutilans* grows on conifer wood and infrequently on other substrates, although the author has collected it on birch (*Betula*) and bracken-fern (*Pteridium*) (Watling, 1984). The distribution of *Fomes fomentarius* in the British Isles is well known being found commonly north of Perth on birch (*Betula*). But to the south it is much rarer, occurring on a range of hosts including maple (*Acer campestre*) and beech (*Fagus sylvatica*) and in southern England almost exclusively on beech (Whalley & Watling, 1982). The climate of western Scotland favours many southern vascular plants which find the central and eastern areas too cold and/or dry. *Fomes fomentarius* while maintaining birch as its main host, in the west of Scotland is found on Mull, and even as far north as Doire Dunn, Lochaber (J. White pers. comm.) on beech. It has also been recorded on *Alnus* from Loch Scresort, (B. J. Coppins, pers. comm.).

HOST-RELATIONSHIPS: ASCOMYCETES
Parallel examples are found in the Ascomycotina: species of *Hypoxylon* were thought to be restricted to particular hosts, but, in a similar way to how one now interprets the associations of mycorrhizal fungi, it is better to consider these xylariaceous fungi as occurring on preferred hosts. Thus, *Hypoxylon fragiforme* is usually found on beech (*Fagus*) but none the less has been recorded occasionally on sycamore (*Acer pseudoplatanus*) and

sweet chestnut (*Castanea sativa*); the closely related *H. cohaerens* on the other hand is less specific and in the British Isles is recorded on a whole range of hosts. *Daldinia concentrica* usually on ash (*Fraxinus*) in England is more commonly found on birch (*Betula*) in Scotland (Whalley & Watling, 1982); in the Hebridean Islands it is rare being only recorded for Laggan Deer Forest, Mull and there on ash!

As records accumulate and more critical studies are completed it is becoming increasingly apparent that in general many of the micro-fungi show much more specific relationships than the familiar larger fungi. The species of the discomycete genus *Dasyscyphus,* unlike those of *Hypoxylon,* exhibit a whole range of specific habitat patterns, particular species colonizing a small range of hosts, or even a single host, e.g. *D. apalus* on *Juncus, D. distinguendus* on *Populus, D. dryinus* on *Betula,* and *D. barbatus* on *Lonicera.* Very few species are found on a wide spectrum of herbaceous substrates but *D. virgineus* may in addition to herbaceous stems grow on woody twigs, trunks and partially worked wood. In parallel to *Dasyscyphus* many pyrenomycetes are also found to be rather specific in their host preferences; see also *Torrendiella eucalypti* below.

Conclusion

British mycologists will for a long time rely heavily on the 'Mull Flora' for a measure of the diversity of fungi found off mainland Scotland. As expansion of our knowledge inevitably depends on interested persons collecting fungi, whenever, and wherever, they can, a bias in favour of the larger more conspicuous members of the flora is always evident. Such a bias is always a stumbling block in this kind of exercise. Whenever a concerted effort is made to collect fungi, especially the less conspicuous microscopic species, several interesting taxa always seem to be found. Indeed the more collecting is carried out in a particular area the more the distribution of selected, even familiar species is extended, and previously thought well-defined often sacrosanct distribution-patterns upset.

The single British record of *Squamanita paradoxa* until comparatively recently was based on a collection from Mull. This agaric was described from N. America and has been subsequently recorded a few times from continental Europe; it has now been collected in Southern England (Reid, 1983). This illustrates our lack of basic information even on the more obvious larger fungi; when the micro-fungi are considered rather amazing distributions are illustrated by the disjunct pattern shown by such oddities as *Martinina panamensis* from South and North America and Rhum (!), *Hobsonia mirabilis* from North and South America and Isle of Ulva (!), *Hypocreopsis rhododendri* from N. America and Mull, Argyll and S. Ireland (!) and *Propolis emarginata* from Australia, Africa, California, and Mull (!). Further work may extend the range of these species and the riddles finally explained; thus the last species has recently been recorded on *Eucalyptus cordata* and *E. gunnii* from Inverewe (Minter, 1983). It is very doubtful whether the distributions of vascular plants demonstrated by

Heslop-Harrison (1952) offer an explanation to these distributions of fungi. Also on *Eucalyptus* on Mull is *Torrendiella eucalypti,* a species previously known from N. Africa and which very probably has been introduced to Mull along with the host-trees to which it is tied. Careful collecting in Australasia will undoubtedly solve the distribution of this fungus on a world-scale.

The apparently incongruous records of the Californian *Calonema luteolum* on Colonsay may be explained differently, viz—a lack of full understanding of the taxonomy and nomenclature of some of the slime-moulds. Thus, Ing (pers. comm.) considers that *C. luteolum* (recorded from Colonsay, Barra and Mingulay) could well be just a form of *Perichaena liceoides,* but he indicates that he requires to see more material; *P. liceoides* is known from Somerset and Germany. It may be that other, at first inexplicable distributions, may be explained by a better understanding of the taxonomy of our organisms.

Taking the total flora, however, there is no evidence of a special fungus flora for Mull and its close islands. Nor at the more limited level is there evidence that the division of the islands into two made by McVean will be supported by the fungi recorded. As shown above fungi basically follow their host (Dennis & Watling, 1983). Thus, birch (*Betula*) occurs in both areas and in both supports the usual inhabitants, e.g. *Piptoporus betulinus, Fomes fomentarius, Inonotus obliquus* and *Amanita crocea.* Equally, the stunted hazel and oak scrub on basalt escarpments in Mull and some of the other islands show similarities between each other, and with the oak woods around Oban etc. and the hazel woods at Bettyhill, Sutherland. More work is necessary before further comparisons can be made between prominent vegetational features in the islands and mainland communities of similar composition for which agaric lists are available.

Acknowledgements

I am particularly grateful to Mr & Mrs M. Clark, who have made their extensive records available to me, and deposited material of critical collections in the Edinburgh Herbarium. I am also grateful to my colleagues A. P. Bennell and B. J. Coppins for their assistance in collecting and/or unravelling the identity of several of the micro-fungi, and especially to Norma Gregory who has helped a great deal with the preparation of the final compilation.

List of Fungi

The general classification adopted in the list follows the entries in the Mull Flora (Henderson & Watling, 1978), as does the format. Where it is felt prudent to diverge from this generalization full author citation is given. Author citation is also given in the entries for all the fungi imperfecti as so many changes have been made in the last few years in both the taxonomy and nomenclature of members of this group. The ascomycetes listed in the Mull List in general followed Dennis (1965); this was not made clear. The same publication has been used herein except where indicated, and then the third edition of the same work (Dennis, 1978) has been relied upon; reference to recent papers including Hebridean records are included after the entry. Generic names have not been used as headings in the list.

Eumycota: Basidiomycotina
Hymenomycetes
AGARICALES

GOMPHIDIACEAE
Gomphidius rutilus — Under conifers, Fishnish Bay. [Better placed in *Chroogomphus*.]

HYGROPHORACEAE
*Hygrocybe ceracea — On graves, Pennygown, 13 x 1969.
*H. obrussea — Kilninian, 16 ix 1980, M. C. Clark.
*H. nitrata — Grassland, Pennygown, 13 x 1969; Loch Buie, 15 x 1967.
H. schulzeri — Graveyard at small church approaching Glen Forsa, 27 x 1970.
H. subviolacea — In short grass by roadside, Glen Forsa, 13 x 1969.
*H. unguinosa — Torosay, 14 x 1969.
*Hygrophorus hypothejus — Under conifers, Salen, 12 x 1969, P. James (E).

'PLEUROTACEAE'
Cheimonophyllum candidissimum (Berk. & Br.) Sing. — On moss and debris, Croggan, 15 ix 1980, M. C. Clark (E). [Placed in *Pleurotellus* in Dennis, Orton & Hora, 1960.]
Leptoglossum retiruge — Amongst moss, Aros Park Wd (E). [See also *Omphalina* below.]
Phaeotellus acerosus — On rocky soil, Kilfinichan, 11 ix 1980. [Placed in *Pleurotellus* in Dennis, Orton & Hora, 1960.]

TRICHOLOMATACEAE
Dermoloma phaeopodium P. D. Orton — In grass on graves, Loch na Keal, x 1969. [A recently described species—see Orton, 1980.]
Fayodia bisphaerigera — Salen Forest, x 1969 (E).
Marasmius undatus — In sandy grassland, Loch Buie, 15 x 1969; Calgary, 11 x 1969, P. James (E).
Mycena adonis — Under *Picea*, Fishnish, 17 x 1980, Marjorie E. Clark (E).
M. crispula — On leaf of *Prunus laurocerasus*, Aros Park, 12 ix 1980, M. C. Clark (E).
M. longiseta — On oak and bracken litter, Salen, vi 1981, M. C. Clark.
M. mauretanica — On wet branch, Torosay, vi 1981, M. C. Clark
M. pudica — On old *Juncus* stems, Croig, M. C. Clark.
*M. rorida — On *Picea* debris, Kintallen, 14 ix 1980, M. C. Clark.
†Omphalina cupulatoides — [The record of *Leptoglossum rickenii* from Fishnish Bay (*Mull Flora* 15.11) is referable to this new species—Orton, 1976a.]
*O. hepatica — Glen Gorm, x 1969.
†O. wynniae — On rotting wood, Torosay, 1969; also probably material from mossy base of trunk, Quinish should be referred here.

† records of interesting species already noted in the Mull Flora
* confirmatory records

Tricholoma album (cf. *pseudoalbum* Bon)	In *Betula/Corylus* wood, 13 ix 1980, M. C. Clark (E).
*T. columbetta	Under *Fagus*, Salen Wood, x 1969.
Tricholoma (Calocybe) gambosum	Iona.

Reduced forms

Calyptella capula (Holmsk.: Fr.) Quél.	On dead herbaceous stems, Kilninian, M. C. Clark (E).
Chromocyphella muscicola (Fr.) Donk	On mossy living trunks of *Quercus* and *Fagus*, Torosay, 17 ix 1980, M. C. Clark (E). [Often assigned to 'Crepidotaceae'.]
Flagelloscypha minutissima (Burt) Donk	On *Juncus* and *Carex*, Meadhouse Loch, iv 1969.

ENTOLOMATACEAE

Eccilia rhodocyclix	Glen Aros.
†Nolanea rhombispora Kühn. & Boursier	In grass, Ulva, 11 ix 1968, Orton (E). [—Orton, 1984.]
N. versatilis	Under clump of bamboo, on damp ground with mosses and liverworts, Torosay Castle Gardens, vi 1981, Marjorie E. Clark.
Leptonia querquedula (Romagn.) Orton	In boggy pasture, Penmore Mill, Orton (E). [—Orton, 1976b.]

CORTINARIACEAE

Cortinarius bibulus	[The record from Aros Forest Park, M. C. Clark (E) refers to *C. lilacinopusillus* P. D. Orton. (Basidiospores 9.5–10 x 5.5–6.5 μm.)—Orton, 1980.]
†C. cinnabarinus	Croggan, 15 ix 1980, M. C. Clark (E).
C. callisteus	Under *Pinus*, Aros Wood, 16 ix 1967.
C. lepidopus	Under *Betula*, An Cairealach, 14 ix 1967.
C. malachoides	Salen Wood, 13 ix 1969.
C. puniceus	Aros Wood, 16 ix 1967.
*C. pseudosalor	Ulva, 18 ix 1967; An Cairealach, 18 ix 1967.
†C. subtriumphans	Under *Betula*, Glen Forsa, 13 ix 1969.
Crepidotus phillipsii	[See *Melanotus* below.]
Galerina cf. praticola	In grassy area, Dishig, 24 x 1970 (E).
*G. mycenopsis	Pennyghael, 9 x 1969; Loch Buie, 15 x 1969; Ulva, 16 x 1969; in moss on sandy shore, Loch Buie, 18 x 1969.
G. nana	On soil, Glen Gorm, 16 x 1969.
G. unicolor	On rotting wood, 11 x 1969.
†Gymnopilus junonius	On dead *Ulex* stems, Loch Buie, 15 x 1969.
Hebeloma anthracophilum	Loch Ba. [No further details.]
*H. sacchariolens	Ulva, 18 ix 1967.
Inocybe (Clypeus) acuta	Under *Fagus*, Aros, 1969, P. James (E).
I. boltonii	Under pine, Salen Forest, 12 x 1969, P. James (E).
I. languinella	Loch Ba. [No further details.]
I. languinosa	Under *Quercus*, Loch Ba, P. James (E).
I. longicystis	No locality, 1969, P. James (E). [No further details.]

I. trechispora	In forestry area, Fishnish, 25 x 1970 (E).
*I. umbrina	[An additional record from Loch Ba, P. James (E).]
*Inocybe (Inocybe) abjecta?	Under conifers, Ulva, 10 x 1969, P. James (E). [See *Mull Flora* 15.15.]
I. cookei	3 collections, one unlabelled: Pennyghael, 9 x 1969, P. James (E); Loch Ba.
I. corydalina	Loch Ba [Has marginal cystidia similar to those of I. *fuscomarginata* Kühn. but lacks crested cystidia.]
I. friesii	Under *Corylus,* Ulva, 10 x 1969.
I. lacera	Under *Fagus,* Loch Buie, 15 x 1969.
I. maculata aff.	In Betula wood, Loch Ba, R. W. G. Dennis. [Differs in size, bright rust-coloured cap and small spores.]
I. microspora	Under conifers and *Corylus,* Ulva, 10 x 1969; two collections, Salen Wood, 13 x 1969, P. James (E).
I. pyriodora	Loch Ba, P. James.
I. subtigrina?	[Without field data this record cannot be confirmed (E).]
I. virgatula	Loch Ba, P. James (E).
I. xanthocephala	Under *Quercus* in field, Loch Ba, P. James.

BOLBITIACEAE

*Agrocybe arvalis	In troop in *Sphagnum,* Loch Ba.
A. paludosa	In boggy area, Dishig.
†Conocybe lenticulospora Watling	On dung, 19 ix 1969, Orton (E). [—Watling, 1980. As 2nd '*Conocybe* species nova' in *Mull Flora* 15. 17.]
†C. murinacea Watling	On dung, Dervaig, 7 ix 1968, Watling (E). [—Watling, 1980. As 1st '*Conocybe* species nova' in *Mull Flora* 15. 17.]

STROPHARIACEAE

*Hypholoma marginatum	In plantation, near Salen, ix 1980, M. C. Clark (E).
Melanotus phillipsii	On wet grass stems, Torosay, 17 ix 1980 and Treshnish Farm, vi 1981, M. C. Clark. [Formerly placed in *Crepidotus.*]

COPRINACEAE

†Coprinus patouillardii var. lipophilus Heim & Romagn.	On grass in corner of sheep pen, possibly associated with old sheep corpse, Dishig. [See *Mull Flora* 15. 18.]
C. triplex P. D. Orton	On cow dung, Gruline, Loch Ba, 25 viii 1973, Orton (E). [Apparently mating tests between this taxon and *C. trisporus* Kemp & Watling suggest the two are conspecific. —Orton, 1976b.]
†Panaeolus castaneifolius, (Murrill) O'lah	*Mull Flora* 15. 18 as *Panaeolina* species nova? [In litt. Michaeli (pers. comm.).]
Psathyrella pseudocasca Romagn.	On rotting pine-stump, Salen Forest, 12 x 1969. [Agrees with Romagnesi's description in Kühner & Romagnesi (1953). Basidiospores 6.5–7.7 x 4.5 μm.]
P. sarcocephala	In clusters at base of *Fagus,* Salen Forest, 13 x 1969, P. James (E).

† records of interesting species already noted in the Mull Flora
* confirmatory records

P. spadicea	Under deciduous trees, Torosay Wd., 14 x 1969, P. James (E).
†P. subnuda	On detritus of dead *Acer pseudoplatanus,* Calgary, 11 x 1969, P. James (E). [A collection differing in broader marginal cystidia and slightly phaseoliform basidiospores.]

LEPIOTACEAE

†Lepiota felina	Under conifers, Salen, 12 x 1960, P. James (E).
L. sistrata	No locality, P. James.

RUSSULACEAE

Russula pectinata	Under *Picea,* Ulva, 13 ix 1968.
*R. puellaris	Carsaig, 10 x 1969.

APHYLLOPHORALES

BANKERACEAE

Phellodon confluens (Pers.) Pouzar	No locality, P. James.
†P. melaleucus (Fr.: Fr.) Karst.	Under *Quercus,* Aros (E). [Erroneously referred to as *P. tomentosus* in *Mull Flora* 15.21.]

CLAVARIACEAE

*Clavaria zollingeri	Edge of lawn, Torosay, 17 ix 1980, M. C. Clark (E).
Macrotyphula juncea (Fr.) Berthier	On leaf of *Prunus laurocerasus,* Aros Park, 12 ix 1980, M. C. Clark. [Formerly placed in *Clavariadelphus.*]
Pistillaria puberula	Fishnish Bay. [Placed in *Typhula* by Berthier, 1976.]
P. erythropus	Glen Forsa, 13 ix 1968. [Placed in *Typhula* by Berthier, 1976.]
*Ramariopsis kunzei	Aros Wood. [No further details.]

CONIOPHORACEAE

Serpula pinastri (Fr.) Bond.	Fishnish Bay: det. J. Ginns.

CORTICIACEAE INCLUDING DICTYONEMATACEAE

*Botryobasidium conspersum J. Erikss.	See *'Oidium'* below.
Cristella submicrospora (Litsch.) Christ.	Aros Wood, 17 v 1968 (E).
Cylindrobasidium evolvens (Fr.) Julich	[As *Corticium* in *Mull Flora* 15.22]
Dictyonema interruptum (Carm.: Hook.) Parm.	Tobermory (BM). [Lichenized resupinate. —Coppins & James (1979).]
Gloeocystidiellum luridum (Bres.) Boid.	On *Fagus,* Tobermory, 18 v 1968; on *Fagus,* Aros Wood, 18 v 1968.
Hyphoderma macedonicum (Litsch.) Donk	On *Fagus?,* Congleton property, Ulva, 15 ix 1968.
*H. pallidum (Bres.) Donk	On *Quercus?,* 19 v 1968.
H. praetermissum (Karst.) Erikss. & Strid	Aros Wood, 18 v 1968.
Laetisaria fuciforme (Berk. & Br.) Burdsall	[As *Corticum* in *Mull Flora* 15.22.]

Mycoacia uda (Fr.) Donk	Torosay.
'Oidium' conspersum (Link) Linder	Aros Wood, 2 collections, 17 v 1968; Glen Forsa, v 1968 (E). [Conidial state of *Botryobasidium*.]
*Peniophora incarnata (Fr.) Karst.	On *Ulex*, Glen Forsa, 21 v 1968, Watling (E), 2 collections; on *Acer* & *Salix* Salen, 30 v 1969, Watling (E).
*P. lycii (Pers.) Höhn. & Litsch.	On *Acer*, Carsaig, 20 v 1968, Watling (E).
Phanerochaete sordida (Karst.) Erikss. & Ryv.	On *Corylus*, Carsaig, 20 v 1968, Watling (E), 2 collections. [Some septate cystidia present.]
P. velutina (Fr.) Karst.	On *Fraxinus*, Carsaig, 20 v 1968, P. James (E).
*Vuilleminia comedens (Nees: Fr.) Maire	Aros Wood, 18 v 1968.

POLYPORACEAE

Antrodia albida	On *Fagus*, Loch Buie, 15 x 1968, P. James (E).
†Fomes fomentarius	On *Fagus*, Ulva, 1968, J. Lee; on *Fagus*, Aros Forest Part, ix 1980, M. C. Clark. [Apparently not yet recorded on birch (?).]
Meripilus giganteus	On *Fagus*, Croig, ix 1980, M. C. Clark.
Polyporus melanopus	Loch Buie (E).
*P. nummularis	On dead wood, Grass Point, M. C. Clark (E).
Tyromyces albellus	On *Fraxinus*, Loch Ba, 8 x 1969.
†T. caesius	On *Fagus* ?, Glen Gorm, 16 x 1969. On grass stems, Quinish, 6 ix 1968 (K). [Agrees with *T. subcaesius* David, which is considered synonymous with *T. caesius* (Pegler in litt., but see Ryvarden, 1976).]
Tyromyces—Ptychogaster albus	Ulva, 10 x 1969.

GANODERMATACEAE

Ganoderma adspersum (S. Schultz) Donk	[Erroneously recorded as *adpressum* in *Mull Flora* 15.23.]

GOMPHACEAE

Ramaria flava	Tobermory, 1910, Cotton. [—Cotton & Wakefield, 1919.]

THELEPHORACEAE

Thelephora anthocephala	Pennyghael, 18 ix 1980, M. C. Clark (E).
†T. terrestris f. resupinata	[Aros Forest Park, 1 ix 1976, M. C. Clark (E). Resembling a *Tomentella* in gross morphology.]
T. palmata	No locality. [—Stevenson, 1882. Omitted from *Mull Flora*.]

'HYMENOMYCETOUS HETEROBASIDIAE'

TREMELLALES

TREMELLACEAE

†Exidia albida	On twigs, All a' Ghael, 24 x 1970; Loch Buie House, 15 x 1969. [Basidiospores broader than generally accepted—Reid, 1970: 420.]

† records of interesting species already noted in the Mull Flora
* confirmatory records

Sebacina calcea	On old wood, Aros Wood, 18 v 1968. [Only a few basidiospores present. Also in Stevenson (1882) as *Corticium.*]
S. podolachia	On effete *Chaetosphaeria,* on ? *Fagus,* Congleton property, Ulva, 13 x 1968.
†Tremella mesenterica	On dead *Salix,* Ulva. [Entirely conidial.]

DACRYMYCETALES

DACRYMYCETACEAE

†Dacrymyces stillatus	On *Ulex* stock, Loch Buie, 19 v 1968; on *Alnus,* 12 ix 1968; on *Picea,* 17 v 1968.
Guepinopsis alpina	On *Pinus sylvestris,* Glen Aros, below Loch Frisa 1971—Reid, 1974. [The first British and European record; this species has subsequently been found in other parts of Scotland.]

GASTEROMYCETES

LYCOPERDALES

LYCOPERDACEAE

Calvatia utriformis	In rough pasture, Port na Ba, vi 1981, M. C. Clark.
*Lycoperdon molle	Under conifers, Knockroy, 12 ix 1968; grassy plot, Drumlairg cottage, 24 x 1970.

HYMENOGASTRALES

HYDNANGIACEAE

Hydnangium careum	Under *Eucalyptus,* Calgary House, M. C. Clark (E) [—Hawker, 1974]; under *Eucalyptus,* Torosay, vi 1981, M. C. Clark.

HEMIBASIDIOMYCETES

UREDINALES

MELAMPSORACEAE

†Melampsora hypericorum	On *Hypericum pulchrum,* nr Dervaig, M. C. Clark (E).
Miyagia pseudosphaeria	On *Sonchus arvensis,* Iona.

PUCCINIACEAE

†Puccinia caricina	On *Ribes uva-crispa* especially fruits, Lettermore (E) and Glen Forsa House, M. C. Clark; on *Carex nigra,* Dervaig, 4 ix 1968 and Salen, 8 ix 1968; on *R. uva-crispa,* 20 v 1968; on *Urtica dioica,* Tornish, 18 v 1968—all D. M. Henderson (E).
*P. festucae	On *Lonicera periclymenum,* Croig, M. C. Clark (E).
P. heraclei	On *Heracleum sphondylium,* Nunnery Garden, Iona, vi 1979, M. C. Clark (E).
P. moliniae	No locality. [—Stevenson, 1882. Omitted from *Mull Flora.*]

USTILAGINALES

Entyloma ficariae	On *Ranunculus ficaria,* Fishnish; Duart Castle (E).
E. microsporum	On *R. repens,* Treshnish (E).

EUMYCOTA: ASCOMYCOTINA
HEMIASCOMYCETES
TAPHRINALES

Taphrina pruni — Deforming fruits of *Prunus spinosa,* Dervaig, vi 1979, M. C. Clark.

EUASCOMYCETES
PLECTASCALES
ELAPHOMYCETACEAE

Elaphomyces muricatus Fr. — Under conifers, Fishnish, iv 1981, M. C. Clark.

HYPOCREALES
HYPOCREACEAE

†Hypocrea rufa — No locality. [—Stevenson, 1882.]

Hypocreopsis rhododendri — Not Croggan as in *Mull Flora* 15.30 but Croig. Also found at Knockan, Quinish, Tobermory and Treshnish Farm—Dennis, 1975.

NECTRIACEAE

Nectriopsis candicans — On old *Trichia floriformis* (Myxomycota), Quinish, 12 vi 1979, M. C. Clark (E).

Pseudonectria jungermanniarum — On living *Lophocolea* (Hepaticae), Quinish and Aros Forest Park.

CLAVICIPITALES
CLAVICIPITACEAE

†Claviceps purpurea — Perithecia on *Molinia caerulea,* Carsaig and Salen, vi 1981, M. C. Clark.

Cordyceps forquignoni — Presumably on dead fly, Quinish, vi 1981, Marjorie E. Clark.

C. gracilis — On pupa, Uisken; on buried pupa, Quinish, vi 1981, M. C. Clark.

SPHAERIALES
DIATRYPACEAE

Diatrype bullata — On *Salix* wood, near Salen.

Endoxyla cirrhosa (Pers.: Fr.) E. Müller & Arx — On cut log, Carsaig (E).

DIAPORTHACEAE

Gnomoniella coryli (Batsch) Sacc. — No locality. [—Stevenson, 1882. Omitted from *Mull Flora.*]

Pharcidia pelvetiae Sutherland — On *Pelvetia caniculata,* 31 v 69, Glen Forsa and Penmore Hill (E). [—J. & E. Kohlmeyer, 1979.]

† records of interesting species already noted in the Mull Flora
* confirmatory records

	LASIOSPHAERIACEAE
Coniochaeta hansenii	Fishnish Bay.
Lasiosphaeria hirsuta	On dead woody stem of *Polygonum,* Aros Forest Park (E).
Podospora pyriformis (Bayer) Cain	On old dung, Fishnish Bay.

	XYLARIACEAE
Anthostomella fuegiana Speg.	Records of *A. tumulosa* on *Luzula* should be referred here. *Mull Flora* 15.33.

PHACIDIALES

	HYPODERMATACEAE
Hypoderma hederae	On *Hedera* leaves, Tobermory and Calgary.
Lophodermium caricinum	On dead *Carex (? sylvatica),* Tobermory.
*L. juniperinum	Gometra. [No further data.]
L. petiolocolum Fuckel	On veins of dead leaves of *Quercus,* Tobermory; Salen, vi 1977, M. C. Clark (E).
Lophomerum ponticum Minter	On *Rhododendron ponticum;* [Recorded as *Lophodermium rhododendri* in *Mull Flora* 15.34]
Phacidium lacerum	On old pine-needles, Fishnish Point.

HELOTIALES

	DERMATACEAE
Actinoscypha muelleri Graddon	On *Carex flacca,* Iona (E).
Belonioscypha culmicola	On small pieces of dead grass, Ledmore, vi 1981, Marjorie E. Clark (E).
Hysteropezizella dowardensis Graddon	On *Carex flacca,* Port na Ba, [—Graddon, 1974.]
H. foecunda	Typical material on *Trichophorum caespitosum.* [Material on *Scirpus caespitosus* from Beinn A'Chailachaidh differed very slightly in microscopic details.]
H. pusilla	On *Juncus effusus,* Tiroran (E); Achronich, vi 1977, M. C. Clark (E).
Leptotrochila brunnellae	On living *Prunella vulgaris,* nr. Ledmore Farm.
L. verrucosa (Wallr.) Schüepp	On living and fading leaves of *Galium saxatile,* Salen, vi 1981; nr. Grass Point, vi 1981; Lagganulva, vi 1981, —all M. C. Clark.
Mollisia junciseda Karst.	Croig, M. C. Clark.
M. maculans	Base of *Nardus stricta* tussock, Scallcastle Bay.
M. melatephra (Lasch) Karst.	On *Carex flacca,* Quinish & Port na Ba (E).
M. millegrana Boud.	On dead stems of *Filipendula,* Quinish and Treshnish, vi 1977, M. C. Clark (E).
M. mutabilis (Berk. & Br.) Massee	On *Deschampsia caespitosa,* Croig, M. C. Clark.
M. rubi	On dead *Rubus,* Loch Buie and Tobermory (E).
Naevia minutissima	On *Quercus* leaves, Tobermory; Lagganulva and Salen, vi 1981, M. C. Clark.
†Niptera phaea	On *Trichophorum caespitosum,* Ross of Mull, viii 1968.
N. poae (Fuckel) Rehm	On grass, Carsaig, M. C. Clark.

Niptera sp.	On *Juncus,* Dervaig, (E).
Pezicula cinnamomea	On bark of *Salix* (? *caprea*), Aros wood (E & K).
P. cf. scoparia	On *Eucalyptus* leaves, Torosay, M. C. Clark (IMI).
Phragmonaevia fuckelii Rehm	On *Peltigera* thalli, Killiemore, Marjorie E. Clark.
Pirottaea brevipila	On dead stems of *Centaurea,* Balliscate (Tobermory).
Pirottaea sp.	On dead *Valeriana,* Glen Forsa, P. James.
*Pseudopeziza trifoli-repentis	[Mull collections of *P. trifolii* should be referred here; *Mull Flora* 15.36.]
Pyrenopeziza foliicola	On decaying *Alnus* leaves, Salen, vi 1981, M. C. Clark (E).
P. lychnidis (Sacc.) Rehm	On dead stems of *Silene,* Torosay Castle.
P. mercurialis (Fuckel) Boud.	On dead stems of *Mercurialis perennis,* Port na Ba.
P. plantaginis Fuckel	On leaves of *Plantago lanceolata,* Ledmore, M. C. Clark.
P. rubi (Fr.) Rehm	On *Rubus idaeus,* Tobermory and Torosay (E).
Tapesia livido-fusca (Fr.) Gill.	On *Rubus fruticosus,* Dishig, Watling (E).
†Trichobelonium melaleucoides	Loch Meadhorn, Watling (E).

HYALOSCYPHACEAE

Dasyscyphus acuum	On *Abies* needles, Glen Forsa Hotel.
D. calyculiformis (Fr.) Rehm	On fallen *Corylus,* Quinish, R. Evans (E).
D. cerinus	On rotten wood, Treshnish (E); on *Quercus* (?), 18 v 1968, Watling (E).
D. crystallinus	On small dead branch of *Quercus,* Portfield, vi 1977, M. C. Clark; Tobermory, vi 1971 (E).
†D. fuscescens	On leaves of *Salix aurita* (E). [Recorded in the *Mull Flora* 15.36 on *Quercus.*]
D. mollissimus	On dead herbaceous stems: *Rumex,* Craignure; *Mercurialis,* Treshnish (E).
D. nudipes var. minor	On dead stems of *Chaemanerion angustifolium,* Aros Forest Park; Treshnish.
D. pudibundus	On dead twig of *Alnus?,* Aros Forest Park (E).
D. rhytismatis (Phill.) Sacc.	On dead leaves of *Vaccinium,* Craignure and Loch Frisa.
Hyaloscypha lachnobrachya (Desm.) Nannf.	Loch Buie, M. C. Clark.
H. laricionis (Vel.) Nannf.	On conifer needles, Glen Aros.
Incrupila melatheja	On *Rubus fruticosus;* —Dennis, 1971.
Lachnellula resinaria (Cooke & Phill.) Rehm	On resinous exudations on *Picea,* Fishnish, vi 1981, Marjorie E. Clark (E).
Lachnellula sp.	On *Alnus* bark, Glen Aros, vi 1979, M. C. Clark. [This is *Trichoscyphella* 'tax. sp. 2' of Dennis (1949).]
Torrendiella eucalypti	On *Eucalyptus* leaves, Calgary and Torosay, M. C. Clark (IMI; E). [Placed in *Zoellneria* in Dennis (1978).]
Urceolella carestiana (Rob.) Dennis	On dead stems of *Athyrium felix-femina,* Quinish, vi 1981, Marjorie E. Clark.
Zoellneria eucalypti	See *Torrendiella eucalypti* above.
Z. rosarum	On fallen leaves of *Rubus,* Aros Forest Park, M. C. Clark (E) [probably also present on *Rosa* at Dervaig and Tiroran.]

† records of interesting species already noted in the Mull Flora
* confirmatory records

	HELOTIACEAE
Allophylaria macrospora (Kitsch.) Nannf.	On decorticated twig of *Salix,* Aros Park (E & K).
Calycellina indumenticola Graddon	On dead leaves of *Salix aurita;* three collections—Graddon, 1974.
C. punctiformis (Grev.) Höhnel	On dead *Quercus* leaf, Gruline (E).
C. spiraeae (Roberge & Desm.) Dennis	On dead *Filipendula,* Treshnish (E).
Chloroscypha seaveri	On dead patches of leafy twigs of *Thuja,* Aros Forest Park.
Echinula asteriadiaformis Graddon	Several localities; type from Glen Aros; on *Rubus fruticosus* (E & K)—Graddon, 1977.
Gorgoniceps aridula	On *Pinus* cones in wet moss, Kintallen nr. Salen (E & K).
Hymenoscyphus repandus	On *Filipendula,* Treshnish, M. C. Clark (E).
H. rhodoleucus	Fishnish Bay (E).
H. rokebyensis Svrček	On *Fagus* cupules, Tobermory, M. C. Clark.
Lanzia stellariae (Vel.) Spooner	On *Stellaria alsine,* Torosay, 15 vi 1981, M. C. Clark (E)—Spooner, 1981.
†Neobulgaria lilacina (Wulf.: Fr.) Dennis	On conifer bark, Lettermore Forest.
Pezizella filicum (Phill.) Sacc.	On dead fern stems, M. C. Clark (E).
P. nigrocorticata Graddon	On dead grass stem, Killiemore, vi 1979, M. C. Clark—Graddon, 1977.
P. roburnea Vel.	On dead *Betula* leaves, Quinish, M. C. Clark (E).
Pezoloma sp.	On *Eriophorum,* Mishnish.
Propolis emarginata (Cooke & Massee) Sherwood	On *Eucalyptus* leaves, Torosay, M. C. Clark (IMI). [First British record of this N. American discomycete.]
Psilachnum tami (Lamy) Dennis	On dead *Mercurialis* stems, Treshnish Farm, vi 1981, M. C. Clark (E).
*Trochila laurocerasi	On dead leaf of *Prunus laurocerasus,* near pier, Tobermory, vi 1981, M. C. Clark.
Tryblidium carestiae	On decayed *Rubus* stem under *Rhododendron,* Aros Park Wood, vi 1981, M. C. Clark. [Typical material has a greyish disc; the Mull collection has a bright orange disc but in no other way differs from *T. carestiae.*]
Vorarlberiga renitens	On algal mat amongst mosses etc., Aros Wood (E).
	OMBROPHILOIDEAE
Claussenomyces canarensis Ouellette & Korf	On twigs, All a' Ghael, x 1970.
	POLYDESMIOIDEAE
Polydesmia pruinosa	On old stroma of *Hypoxylon multiforme* (Pyrenomycete), Loch Ba, vi 1977, M. C. Clark.
	SCLEROTINIACEAE
Ciboriopsis simulata (Ellis) Dennis	On decaying leaves of *Rubus,* Grasspoint, Marjorie E. Clark. [Placed in *Moellerodiscus* by Dumont, 1976.]
Rutstroemia conformata	On decaying leaves of *Alnus,* Aros Forest Park and Killiechronan, vi 1977, M. C. Clark (E).
R. fruticeti	On dead canes of *Rubus fruticosus,* Killiemore.

R. hercynica	On dead stems of *Chamanerion angustifolium*, Aros Park Wood, vi 1981, M. C. Clark (E).
R. lindaviana	On dead grass, Salen, vi 1981, Marjorie E. Clark (E).
R. petiolorum	On *Fagus* petiole, Gruline, M. C. Clark (E).
†Sclerotinia eleocharidis D. Hend.	[Considered a *Myriosclerotinia* by Kohn (1979) close to *M. sulcata* Whetzel—*Mull Flora* 15: 38.]
S. sclerotiorum	On leaf litter, Tobermory.
S. tuberosa	On tubers of *Anemone nemorosa*, Dervaig. [Now placed in *Dumontia*—Kohn, 1979.]

GEOGLOSSACEAE

†Geoglossum glutinosum	? Loch Ba. [No further details.]
†G. nigritum	*Mull Flora* 15.38. [The record of *Trichoglossum walteri* should be included here.]
Heyderia abietis	On *Picea* needles, Loch Frisa, M. C. Clark (E & K).

LECANORALES

†Dactylospora stygia (Berk. & Curt.) Hafellner	Carsaig, v 1968, P. James; Kilninian, viii 1976, M. C. Clark (E). [Formerly placed in *Buellia* or *Karschia*. Record of *Karschia nigerrima* should be referred here. —*Mull Flora* 15.39 and Hafellner, 1979.]
†Rhizodiscina lignyota (Fr.) Hafellner	Glen Forsa House, v 1968, P. James; Torosay, ix 1968, P. James. [Should include record of *Karschia* sp., Henderson 9001 (E)—*Mull Flora* 15.39.]

OSTROPALES

Apostemidium torrenticola Graddon	On wet wood in streams, Tobermory, vi 1977, M. C. Clark (E).

PEZIZALES

ASCOBOLACEAE

ASCOBOLEAE

Ascobolus brassicae	On mouse droppings, Grass Point, Loch Buie and Loch Don Head (E); on sheep droppings, Salen, vi 1981.
A. carbonarius	On bonfire site, Quinish.
A. immersus	On dung, Fishnish. [Often placed in *Dasyobolus*.]
A. viridis	On bare soil, Tobermory.
Pyronema omphalodes	On bonfire site, M. C. Clark.
Saccobolus citrinus Boud. & Torrent.	Scallastle Bay, M. C. Clark.
S. versicolor (Karst.) Karst.	On rabbit pellet, Aros Park (E); on sheep droppings, Salen, vi 1981, M. C. Clark.

PSEUDOASCOBOLEAE

Ascozonus woolhopensis	On mouse droppings, Scallastle Bay, M. C. Clark.

THELEBOLACEAE

*Thelebolus nanus Heimerl.	On sheep droppings, Salen vi 1981, M. C. Clark.

HELVELLACEAE

Gyromitra esculenta	Salen, R. Evans.

† records of interesting species already noted in the Mull Flora
* confirmatory records

	HUMARIACEAE
Anthracobia macrocystis	Bonfire site, Pennyghael, M. C. Clark (E).
Cheilymenia stercorea	No locality. [—Stevenson, 1882. Omitted from *Mull Flora.*]
Lamprospora crec'hqueraultii	On damp ground covered by film of moss, Loch Don, M. C. Clark (E).
Leucoscypha erminea Bomm. & Roussl.	On litter inside damp *Vaccinium* bush, Loch Frisa.
Neottiella vivida	On sawdust, Glen Gorm, x 1969, P. James; on sand and moss, Loch Buie, x 1969 (E).
Pulvinula constellatio	On twigs, Aros Forest Park (E).
Scutellinia trechispora	No locality. [—Stevenson, 1882. Omitted from *Mull Flora.*]
Sepultaria arenosa	Edge of cinder covered car park, Aros Park, M. C. Clark (E).
	PEZIZACEAE
Otidea onotica	On ground, Quinish, M. C. Clark (E).
Peziza bovina	On old cow dung, Grass Point and Balliscate.
P. echinospora	Bonfire site, Fishnish Point.
Plicaria trachycarpa	Bonfire site, Quinish.
	SARCOSCYPHACEAE
Sarcoscypha coccinea	Tobermory, J. Whittaker.

LOCULOASCOMYCETES

PLEOSPORALES

	PLEOSPORACEAE
†Helminthosphaeria clavariarum	See *Spadicoides* (Deuteromycotina).
Leptosphaeria derasa	On dead stems of *Senecio jacobaea,* Glen Aros and Loch Buie.
Ophiobolus cirsii	On *Cirsium palustre,* Loch Peallach.
	LOPHIOSTOMATACEAE
Lophiostoma pileatum (Tode: Fr.) Fuckel	On bark of tree stump, Salen, R. Evans.

HYSTERIALES

	PATELLARIACEAE
Abrothallus parmeliarum	On *Sticta fuliginosa* (Lichen), on *Fraxinus,* Glen Forsa House.
	ACROSPERMATACEAE
Oomyces carneoalbus	On dead *Deschampsia caespitosa,* three localities, M. C. Clark.
	CHAETOTHYRIACEAE
Chaetothyrium babingtonii	On living *Rhododendron* leaves, Aros Forest Park.

DOTHIDEALES

PSEUDOSPHAERIACEAE

Extrawettsteinia pinastri Barr — On pine-needles, Glen Aros.

MYCOSPHAERELLACEAE

Mycosphaerella ascophylii Cotton — On *Ascophyllum nodosum*, Fishnish Bay, [Immature].

M. clymenia (Sacc.) Oud. — On leaves of *Lonicera*, Quinish.

†M. killiani — [Records of *Cymadothea trifolii* should be referred here. *Mull Flora* 15.43.]

Stigmidium peltidae (Vainio) R. Sant. — On *Peltigera canina*, Lagganulva.

PLEOSPORACEAE

Naumovia abundans — On stems of *Prunella vulgaris*, Aros Forest Park and Quinish, M. C. Clark.

MICROTHYRIALES

MICROTHYRIACEAE

Asterina festucae — On *Deschampsia caespitosa*, Croig.

Aulographina eucalypti (Cooke & Mass.) von Arx & Müller — On leaves of *Eucalyptus*, M. C. Clark (IMI) [second British record].

†Aulographum vagum — Several records. [Interestingly a member of the phylloplane on old leaves of both *Eucalyptus* and *Rhododendron*.]

Echidnodes aulographioides — On twigs of living *Rhododendron*, Aros Forest Park, iv 1977, M. C. Clark. [Often placed in *Lembosina*.]

Microthyrium cytisi var. ulicis J. P. Ellis — On dead *Ulex europaeus*, nr. Salen.

M. microscopicum Desm. — On old leaves of *Quercus*, Croig (E).

†Morenoina clarkii J. P. Ellis — On *Rubus fruticosus*, Aros Park Forest and Glen Aros, M. C. Clark (E). [As *Morenonina* sp. in *Mull Flora* 15.43.]

M. rhododendri J. P. Ellis — On dead twigs of *Rhododendron ponticum* and *Vaccinium myrtillus*. —Ellis, 1980.

Stomiopeltis cupressicola J. P. Ellis — On *Cupressus*, Tobermory. —Ellis 1977b.

S. pinastri (Fuckel) von Arx — On needles of *Abies*, Glen Aros [=*Calothyrium* in Dennis, 1978].

Trichothyrina alpestris (Sacc.) Petrak — On *Eucalyptus* leaves, Torosay, M. C. Clark (IMI). [Usually grows on grasses and sedges.]

T. fimbriata J. P. Ellis — On *Cupressus* leaves, Tobermory—Ellis, 1977a

EUMYCOTA: DEUTEROMYCOTINA
'COELOMYCETES'

Asteromella sp. — State of *Mycosphaerella killiani*: see above.

Ceuthospora lauri (Grev.) Grev. — On *Eucalyptus*, M. C. Clark (IMI) [doubtfully distinct from taxon recorded under *C. laurocerasi* in *Mull Flora* 15.45. —Stevenson, 1882.]

† records of interesting species already noted in the Mull Flora

†C. phacidioides Grev. — On *Eucalyptus,* Torosay, M. C. Clark. [Recorded on other substrates in *Mull Flora* 15.45.]
Chaetoconis polygoni (Ell. & Ev.) Clements — On *Polygonum,* Salen, 21 v 1960, P. James.
Coleophoma cylindrospora (Desm.) Höhnel — On *Eucalyptus,* Torosay (IMI).
Colletotrichum gloeosporioides (Penz.) Sacc. — On fruits of *Prunus spinosa,* Quinish House.
Cytospora eucalypticola van der Westhuizen — On dead leaves of *Eucalyptus,* Torosay, M. C. Clark (IMI) [first British record].
Leptothyrium vulgare Sacc. — On *Lythrum,* Glen Forsa, 21 v 1968, P. James.
Phoma elongata Desm. — Fishnish Bay.
P. macrocapsa Trail — On *Mercurialis,* Salen ? (E).
Phomopsis cf. eucalypti Zerova — Torosay, M. C. Clark (IMI).
Pilidium acerinum Kunze — On dead *Eucalyptus* leaves, Torosay (IMI).
P. concavum (Desm.) Höhnel — On dead *Eucalyptus* leaves, Torosay (IMI).
Readeriella mirabilis H. & P. Syd. — On *Eucalyptus* leaves, Torosay, M. C. Clark (IMI). [Third British record.]
Septoria castaneicola Desm. — On *Castanea* leaves, Torosay, R. W. G. Dennis.
Strasseria geniculata (Berk. & Br.) Höhnel — On *Eucalyptus* leaves, Torosay, M. C. Clark (IMI).

'HYPHOMYCETES'

Arthrinium sporophleum Kunze — On dead *Carex flacca* leaves, Lagganulva, iv 1975, M. C. Clark.
A. euphorbiae M. B. Ellis — On *Ammophila,* Loch Buie, 1977, M. C. Clark (IMI). [First British record; previously only known from Africa.]
Bactridium flavum Kunze: Fr. — On woody debris in stony ground, Aros Forest Park, 12 ix 1980.
†Blistum tomentosum (Schrader: Fr.) Sutton — On old *Trichia botrytis* (Myxomycota), Fishnish Bay, iv 1979, M. C. Clark. [—*Mull Flora* 15.52 under *Tilachlidium.*]
*Camarosporium stephensii Sacc. — On *Pteridium,* Aros Forest Park; two collections, M. C. Clark (E).
Coremiella cubispora (Berk. & Curt.) M. B. Ellis — On canes of *Rubus,* Aros Forest Park, M. C. Clark (E).
Endophragmia alternata Tubaki & Saito — On dead leaves of *Quercus,* Torosay and Uisken, M. C. Clark.
*E. atra (Berk. & Br.) M. B. Ellis — Uisken, M. C. Clark. [No locality given in *Mull Flora* 15.46.]
Epicoccum purpurascens Ehrenb. — Grass Point, M. C. Clark.
Ovularia rufibasis (Berk. & Br.) Mass. — Presumably on *Myrica gale.* —Stevenson (1882).
Gibellula pleiopus (Vuill.) Mains — On spider, Loch Don, M. C. Clark (E).
Isthmolongispora minima Matsushima — On *Eucalyptus* leaves, Torosay Castle Gardens, vi 1981, M. C. Clark. [Known previously only from Arran.]
Polyscytalum hareae (Sutton) Kirk — On dead *Eucalyptus* leaves, Torosay, M. C. Clark (IMI). [Described by Sutton (1976) as *Subulispora.*]

P. truncatum Sutton & Hodges	On dead *Eucalyptus* leaves, Torosay, M. C. Clark.
Ramularia peltigericola Hawksworth	On *Peltigera* thallus, Killiemore, Marjorie E. Clark.
Spadicoides clavariarum (Desm.) Hughes	On *Clavulina* sp. (? *rugosa*), Kintallen near Salen, M. C. Clark (E).
†**Xylohypha nigrescens** (Pers.: Fr.) Mason	As *Torula pulveracea*—Stevenson, 1882.

'PHYCOMYCETES'

EUMYCOTA: ZYGOMYCOTINA

ZYGOMYCETES

MUCORALES

MUCORACEAE

Mortierella pilulifera van Tiegh.	On rodent jaw bone, Carsaig, 10 v 1968.
†**Pilobolus crystallinus** (Tode) van Tiegh.	On dung, An Coire, Loch Spelve.

CHYTRIDIOMYCETES

CHYTRIDIALES

SYNCHYTRIACEAE

Synchytrium mercurialis (Lib.) Fuckel	On *Mercurialis perennis*, Quinish, vi 1979 and Treshnish Farm, vi 1981, M. C. Clark. [Also in Stevenson, 1882.]
S. succisae de Bary & Woronin	On *Succisa pratensis*, ix 1980, Croggan, M. C. Clark.

EUMYCOTA: MASTIGOMYCOTINA

OOMYCETES

PERONOSPORALES

PERONOSPORACEAE

Peronospora conferta (Ung.) Gaumann	On *Cerastium holosteoides*, Kilchrenan.
†**Plasmopara angelicae** (Casp.) Trotter	[Records of *P. nivea* on *Angelica sylvestris* refer here—Mull Flora 15.54.]

MYXOMYCOTA

MYXOMYCETES

LICEALES

LICEACEAE

Licea deplorata	On birch litter, M. C. Clark.

† records of interesting species already noted in the Mull Flora
* confirmatory records

TRICHIALES

TRICHIACEAE

Perichaena pedata   In moist chamber culture; substrate from Loch Buie.

Note: It should be noted that when the proofs of the 'Mull Flora' were pasted-up, the entries for *Reticularia* (p. 15.54) to *Cribraria persoonii* which should have been placed after *Lycogala* (Liceales) were erroneously incorporated into the Trichiaceae. *Arcyria cinerea*—*Perichaena chrysosperma* should close 15.54.

STEMONITALES

STEMONITACEAE

Paradiacheopsis fimbriata   In moist chamber culture; substrate from Calgary House.
Stemonitis splendens   Glen Aros; Croig.

PHYSARALES

PHYSARACEAE

Cienkowskia reticulata   Gruline, ix 1980, Marjorie E. Clark.
Physarum robustum   Loch Buie.
†P. scoticum Ing   Aros Forest Park; type locality—Ing, 1982. [As '*Physarum species nova*' in *Mull Flora* 15.56.]

DIDYMIACEAE
(as Didmyiaceae in *Mull Flora* 15.56)

Didymium minus   On *Prunus laurocerasus,* Aros Forest Park, 12 ix 1980, M. C. Clark.
Mucilago crustacea   Uisken: Iona.

Rejected Names

The following taxa listed for Mull by Stevenson (1882) are to be considered of doubtful position mainly because of difficulties in interpreting the name.
They include:

BASIDIOMYCOTINA:

Agaricus albo-atrus Bolt.   Omitted by Dennis, Orton & Hora, 1960. Traditionally considered to be a *Mycena*.

Agaricus scaber Müll.   Possibly the same as *Inocybe corydalina* Quél.

Lactarius insulsus (Fr.) Fr.   Confused and without herbarium material impossible to place (Rayner, pers. comm.)

Russula rubra s. Cooke, auct. brit.   Probably *R. atropurpurea*—*Mull Flora* 15.20.

Polyporus vulgaris Fr.   A nomen ambiguum.

ASCOMYCOTINA:

Erysiphe martii Lév.   Broad concept; the identity of the Mull collection rests on a knowledge of the host-plant.

Helotium claroflavum   Probably a young state of *Bisporella citrina*—Dennis, 1955; *Mull Flora* 15.38.

H. pallescens   Could possibly be *Bisporella monilifera* (Fuckel) Sacc.— Dennis, 1955.

† records of interesting species already noted in the Mull Flora

REFERENCES

BERTHIER, J. (1976). Monographie des Typhula Fr., Pistillaria Fr., et genre voisins. *Bull. Soc. Linn. Lyon* 45, suppl.: 1–213.
CLARK, M. C. (ed.) (1980). *A fungus flora of Warwickshire*. London.
COPPINS, B. J. & JAMES, P. (1979). Dictyonema interruptum. *Lichenologist* 11: 103–105.
COTTON, A. D. & WAKEFIELD, E. M. (1919). A revision of the British Clavariae. *Trans. Brit. Mycol. Soc.* 6: 164–198.
CURRIE, A. (1979). The vegetation of the Outer Hebrides. *Proc. Roy. Soc. Edinb.* 77B: 219–265.
—— & MURRAY, C. W. (1983). Flora and vegetation of the Inner Hebrides. In BOYD, J. M. & BOWES, D. R. (eds), The Natural Environment of the Inner Hebrides. *Proc. Roy. Soc. Edinb.* 83B: 293–318.
DENNIS, R. W. G. (1949). A revision of the British Hyaloscyphaceae with notes on related European species. *Mycol. Papers* 32: 1–97.
—— (1952). Contribution towards a fungus flora of the small Isles of Inverness. *Trans. Bot. Soc. Edinb.* 36: 58–70.
—— (1955). A revision of the British Helotiaceae in the herbarium of the Royal Botanic Gardens, Kew with notes on related European species. *Mycol. Papers* 62: 1–216.
—— (1964). The fungi of the Isle of Rhum. *Kew Bull.* 19: 77–131.
—— (1965). *British Ascomycetes*. Lehre.
—— (1971). New or interesting British microfungi. *Kew Bull.* 25: 335–374.
—— (1973). The fungi of southeast England. *Kew Bull.* 28: 133–139.
—— (1975). New or interesting British microfungi. III. *Kew Bull.* 30: 345–365.
—— (1978). *British Ascomycetes*, 3rd ed., Vaduz.
——, ORTON, P. D. & HORA, F. B. (1960). New checklist of British agarics and boleti. *Trans. Brit. Mycol. Soc.* 43, Suppl.
—— & WATLING, R. (1983). Fungi in the Inner Hebrides. In BOYD, J. M. & BOWES, D. R. (eds), The Natural Environment of the Inner Hebrides. *Proc. Roy. Soc. Edinb.* 83B: 415–429.
DUMONT, K. P. (1976). Sclerotiniaceae XI. On Moellerodiscus (=Ciboriopsis). *Mycologia* 68: 233–267.
ELLIS, J. P. (1977a). The genera Trichothyrina and Actinopeltis in Britain. *Trans. Brit. Mycol. Soc.* 68: 145–155.
—— (1977b). The genus Stomiopeltis in Britain. *Trans. Brit. Mycol. Soc.* 68: 157–159.
—— (1980). The genus Morenoina in Britain. *Trans. Brit. Mycol. Soc.* 74: 297–307.
GRADDON, W. D. (1974). Some new discomycete species. *Trans. Brit. Mycol. Soc.* 63: 485–495.
—— (1977). Some new discomycete species: 4. *Trans. Brit. Mycol. Soc.* 69: 255–273.
HAFFELNER, J. (1979). Karschia. Revision einer sammelgattung an der Grenze von lichenisierten und nichtlichenisierten Ascomyceten. *Biblio. mycol.* 62. Vaduz.
HAWKER, L. (1974). Revised annotated list of British hypogeous fungi. *Trans. Brit. Mycol. Soc.* 62: 67–76.
HENDERSON, D. M. & WATLING, R. (1978). Fungi in JERMY, A. C. & CRABBE, J. A. (eds.), *The Island of Mull*. London.
HESLOP-HARRISON, J. (1952). Occurrence of the American pondweed Potamogeton epihydrus Raf. in the Hebrides. *Nature* 169: 548.
ING, B. (1982). Notes on myxomycetes III. *Trans. Brit. Mycol. Soc.* 78: 439–446.
KIRK, P. M. & SPOONER, B. M. (1984). An account of the fungi of Arran, Gigha and Kintyre. *Kew Bull.* 38: 503–597.
KOHLMEYER, J. & KOHLMEYER, E. (1979). *Marine mycology. The higher fungi*. London.
KOHN, L. (1979). A monographic revision of the genus Sclerotinia. *Mycotaxon* 9: 365–444.
KÜHNER, R. & ROMAGNESI, H. (1953). *Flore analytique des Champignons supérieurs de France*. Paris.
MASSEE, G. & CROSSLAND, C. (1905). *The fungus flora of Yorkshire*. London.
MCVEAN, D. N. (1964). Regional pattern of the vegetation. In BURNETT, J. H. (ed.) *The vegetation of Scotland*. Edinburgh.
—— & RATCLIFFE, D. A. (1962). *Plant communities of the Scottish Highlands*. London.

Minter, D. W. (1983). Upland foray, Wester Ross. *Bull. Brit. Mycol. Soc.* 17: 34–43.
Moeller, F. H. (1945). *Fungi of the Faeroes.* Copenhagen.
Orton, P. D. (1976a). Notes on British agarics V. *Kew Bull.* 31: 709–721.
────── (1976b). Notes on British agarics VI. *Notes RBG Edinb.* 35: 147–153.
────── (1980). Notes on British agarics VII. *Notes RBG Edinb.* 38: 315–330.
────── (1984). Notes on British agarics VIII. *Notes RBG Edinb.* 41: 565–624.
Reid, D. A. (1970). New or interesting records of British Hymenomycetes, IV. *Trans. Brit. Mycol. Soc.* 55: 413–441.
────── (1974). A monograph of the British Dacrymycetales. *Trans. Brit. Mycol. Soc.* 63: 433–494.
────── (1983). A second British collection of Squamanita paradoxa. *Bull. Brit. Mycol. Soc.* 17: 111–113.
Ryvarden, L. (1976). *The polyporaceae of North Europe.* Oslo.
Spooner, B. M. (1981). New records and species of British microfungi. *Trans. Brit. Mycol. Soc.* 76: 265–301.
Stevenson, J. (1882). Mycologia Scotica: corrigenda and additions. *Scot. Nat.* 6: 213–221.
Sutton, B. C. (1976). Three new hyphomycetes from Britain. *Trans. Brit. Mycol. Soc.* 71: 161–171.
Watling, R. (1962). The larger fungi of the Garth area. *Report Scottish Field Studies Assoc.* (1962): 15–26.
────── (1970). Checklist of the plants of Rhum, Inner Hebrides. Part III. Fungi. *Trans. Bot. Soc. Edinb.* 49: 497–535.
────── (1976). Notes on the fungal flora of Sutherland. In Kenworthy, B. J. (ed.), John Anthony's *Flora of Sutherland.* Edinburgh.
────── (1980). Observations on the Bolbitiaceae 20. New British species of Conocybe. *Notes RBG Edinb.* 38: 345–356.
────── (1981) Relationships between macromycetes and the development of higher plant communities. In Wicklow, D. T. & Caroll, G. C. (eds.), *The Fungal Community.* New York.
────── (1983a). Additions to the fungus flora of the Hebrides *Trans. Bot. Soc. Edinb.* 44: 127–138.
────── (1983b). Fungi of Skye. *Glasgow Naturalist* 20: 269–311.
────── (1984). Larger fungi of birchwoods. In Henderson, D. M. & Mann, D. A. (eds), Birches. *Proc. Roy. Soc. Edinb.* 85B: 129–140.
────── & Richardson, M. J. (1970). Fungi of Loch Druidibeg Nature Reserve. *Bull. Brit. Mycol. Soc.* 4: 97–99.
────── & ────── (1971). Agarics of St Kilda. *Trans. Bot. Soc. Edinb.* 41: 165–187.
Whalley, A. J. S. & Watling, R. (1982). Distribution of Daldinia concentrica in the British Isles. *Trans. Brit. Mycol. Soc.* 78: 47–53.

Index to Genera

Abrothallus, 24
Actinoscypha, 20
Agrocybe, 15
Allophylaria, 22
Anthostomella, 20
Anthracobia, 24
Antrodia, 17
Apostemidium, 23
Arthrinium, 26
Ascobolus, 23
Ascozonus, 23
Asterina, 25
Asteromella, 25
Aulographina, 25
Aulographum, 25

Bactridium, 26
Belonioscypha, 20
Blistum, 26
Botryobasidium, 16

Calocybe, 14
Calvatia, 18
Calycellina, 22
Calyptella, 14
Camarosporium, 26
Ceuthospora, 25
Chaetoconis, 26
Chaetothyrium, 24
Cheilymenia, 24
Cheimonophyllum, 13
Chloroscypha, 22
Chromocyphella, 14
Ciborisopsis, 22
Cienkowskia, 28
Claussenomyces, 22
Clavaria, 16
Claviceps, 19
Coleophoma, 26
Colletotrichum, 26
Coniochaeta, 20
Conocybe, 15
Coprinus, 15
Cordyceps, 19
Coremiella, 26
Cortinarius, 14
Crepidotus, 14
Cristella, 16
Cylindrobasidium, 16
Cytospora, 26

Dacrymyces, 18
Dactylospora, 23
Dasyscyphus, 21
Dermoloma, 13
Diatrype, 19
Dictyonema, 16
Didymium, 28

Eccilia, 14
Echinodes, 25
Echinula, 22
Elaphomyces, 19
Endophragmia, 26
Endoxyla, 19
Entyloma, 19
Epicoccum, 26
Exidia, 17
Extrawettsteinia, 25

Fayodia, 13
Flagelloscypha, 14
Fomes, 17

Galerina, 14
Ganoderma, 17
Geoglossum, 23
Gibellula, 26
Gloeocystidiellum, 16
Gnomoniella, 19
Gomphidius, 13
Gorgoniceps, 22
Guepinopsis, 18
Gymnopilus, 14
Gyromitra, 23

Hebeloma, 14
Helminthosphaeria, 24
Heyderia, 23
Hyaloscypha, 21
Hydnangium, 18
Hygrocybe, 13
Hygrophorus, 13
Hymenoscyphus, 22
Hyphoderma, 16
Hypholoma, 15
Hypocrea, 19
Hypocreopsis, 19
Hypoderma, 20
Hysteropezizella, 20

Incrupila, 21
Inocybe, 14
Isthmolongispora, 26

Lachnellula, 21
Laetisaria, 16
Lamprospora, 24
Lanzia, 22
Lasiosphaeria, 20
Lepiota, 16
Leptoglossum, 13
Leptonia, 14
Leptosphaeria, 24
Leptothyrium, 26
Leptotrochila, 20

Leucoscypha, 24
Licea, 27
Lophiostoma, 24
Lophodermium, 20
Lophomerum, 20
Lycoperdon, 18

Macrotyphula, 16
Marasmius, 13
Melampsora, 18
Melanotus, 15
Meripilus, 17
Microthyrium, 25
Miyagia, 18
Mollisia, 20
Morenoina, 25
Mortierella, 27
Mucilago, 28
Mycena, 13
Mycoacia, 17
Mycosphaerella, 25

Naevia, 20
Naumovia, 25
Nectriopsis, 19
Neobulgaria, 22
Neottiella, 24
Niptera, 20
Nolanea, 14

Oidium, 17
Omphalina, 13
Oomyces, 24
Ophiobolus, 24
Otidea, 24
Ovularia, 26

Panaeolus, 15
Paradiacheopsis, 28
Peniophora, 17
Perichaena, 28
Peronospora, 27
Pezicula, 21
Peziza, 24
Pezizella, 22
Pezoloma, 22
Phacidium, 20
Phaeotellus, 13
Phanerochaete, 17
Pharcidia, 19
Phellodon, 16
Phoma, 26
Phomopsis, 26
Phragmonaevia, 21
Physarum, 28
Pilidium, 26
Pilobilus, 27
Pirottaea, 21
Pistillaria, 16

Plasmopara, 27
Plicaria, 24
Podospora, 20
Polydesmia, 22
Polyporus, 17
Polyscytalum, 26
Propolis, 22
Psathyrella, 15
Pseudonectria, 19
Pseudopeziza, 21
Psilachnum, 22
Ptychogaster, 17
Puccinia, 18
Pulvinula, 24
Pyrenopeziza, 21
Pyronema, 23

Ramaria, 17
Ramariopsis, 16
Ramularia, 27
Readeriella, 26
Rhizodiscina, 23
Russula, 16
Rutstroemia, 22

Saccobolus, 23
Sarcoscypha, 24
Sclerotinia, 23
Scutellinia, 24
Sebacina, 18
Septoria, 26
Sepultaria, 24
Serpula, 16
Spadicoides, 27
Stemonitis, 28
Stigmidium, 25
Stomiopeltis, 25
Strasseria, 26
Synchytrium, 27

Tapesia, 21
Taphrina, 19
Thelebolus, 23
Thelephora, 17
Torrendiella, 21
Tremella, 18
Trichobelonium, 21
Tricholoma, 14
Trichothyrina, 25
Trochila, 22
Tryblidium, 22
Tyromyces, 17

Urceolella, 21

Vorarlbergia, 22
Vuilleminia, 17

Xylohypha, 27

Zoellneria, 21